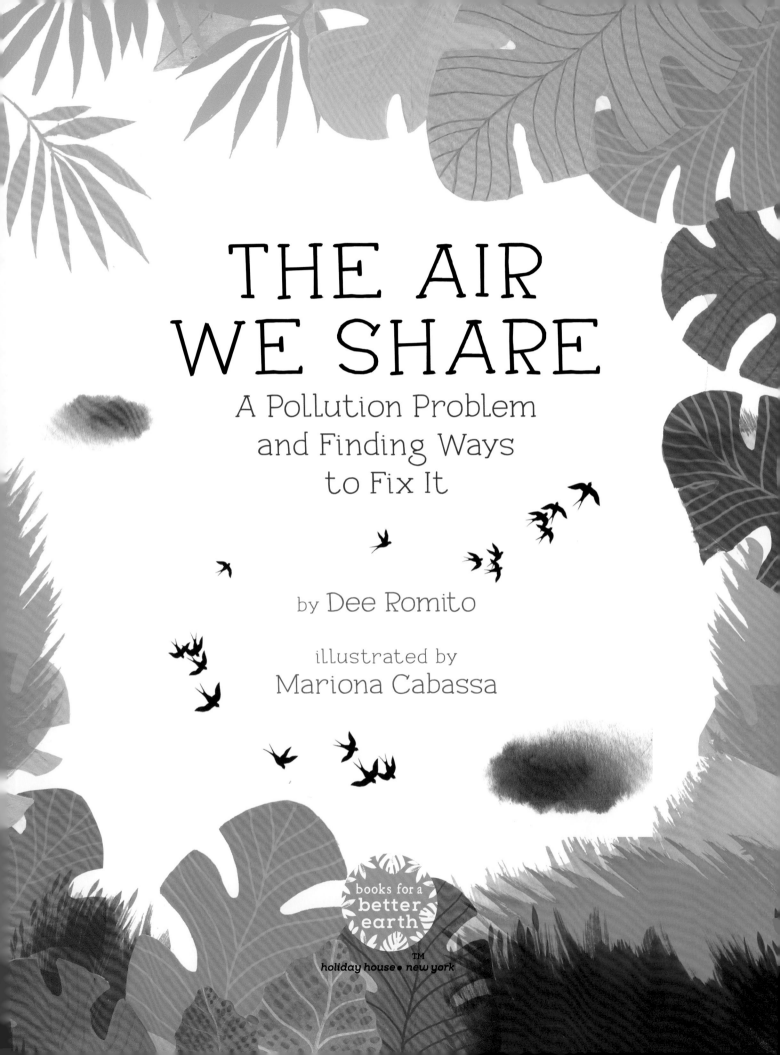

THE AIR WE SHARE

A Pollution Problem and Finding Ways to Fix It

by Dee Romito

illustrated by Mariona Cabassa

books for a better earth™

holiday house • new york

What goes up here,
ends up there.
It's all connected,
in the air we share.

The air we breathe is a mixture of gases and tiny particles. It may seem like those little particles are only a small amount, but they can make a BIG difference.

KRAKATOA

In a place called the Sunda Strait, a volcano named Krakatoa had been sputtering for months. But in August of 1883, things were about to change.

The volcano erupted, shooting clouds of ash up, up, up, high into the air. The area was in darkness for two and a half days, and the haze from the explosion circled the Earth multiple times.

Because of all the gases and ash particles that entered the atmosphere, a lot of sunlight was blocked from reaching the Earth, making it colder all around the globe!

The Earth itself had changed the air around it.

Scientists studied the event and learned everything they could.

THE DUST BOWL

In the 1930s, dust storms blew through the United States across parts of Texas, Oklahoma, Kansas, Colorado, and New Mexico. Too many farmers had moved in to use the land for a kind of farming it was not meant for.

There was also a severe shortage of rain, and the dry soil was no longer held down by the plants that used to be there. The fine particles were carried up, up, up by the strong winds.

Other areas of the country could tell exactly where the dust came from based on its color! The storms carried dust all the way to the Atlantic Ocean.

This all led to one of the worst dust storms in American history on April 14, 1935, when a black wall of dust turned a sunny day into total darkness. They called the devastating storm "Black Sunday."

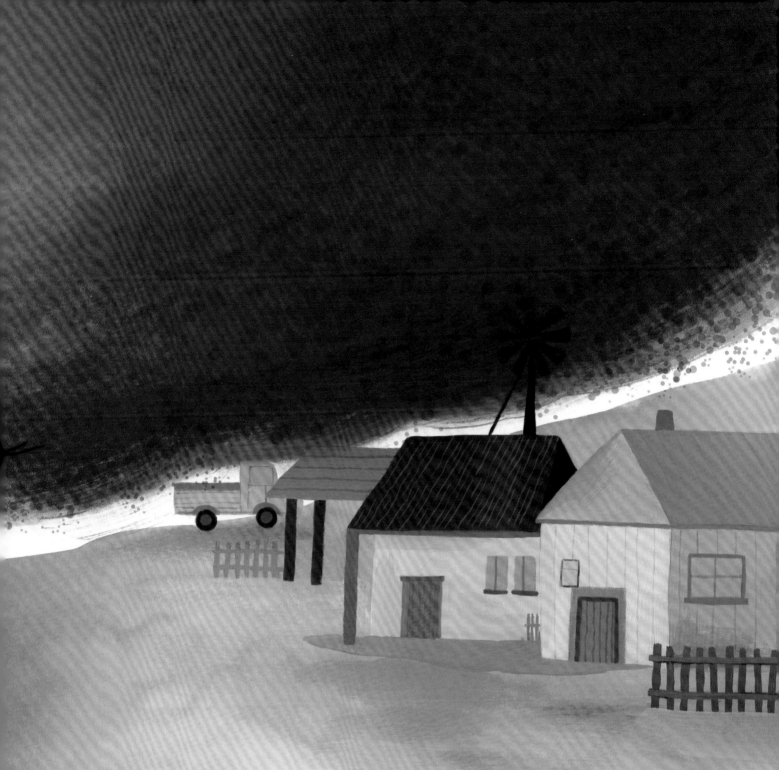

Humans had changed the air around them.

The United States government set up programs to improve the situation. They planted millions and millions of trees, made water systems better, and tried to educate people so they would understand why the dust storms had happened.

THE GREAT SMOG OF LONDON

In 1950s London, people burned coal to heat their homes, and factories used it for power, sending pollutants up, up, up into the air.

Then one day, a dense fog settled in, covering the city. It combined with soot in the air, and when a layer of warm air trapped it all near the ground, it caused The Great Smog of 1952, darkening the sky for five days and making Londoners sick. Thousands of people died as a result.

Pollution had changed the air around the city.

The British Parliament passed the Clean Air Act of 1956, which controlled the burning of coal in homes, regulated the smoke from factories, and led the way in using other sources of energy.

Events like these have shown us how important clean air is and that having a healthy environment makes way for the good things to travel through the air.

Seeds . . .

Wind transports seeds as they float or spin in the breeze. And birds eat seeds and spread them around as they travel, along with nutrients for the soil.

How do birds spread seeds and nutrients? Well, scientists call it guano, but you probably know it as bird poop.

Scent...

There's a factory in Buffalo, New York, that makes the air smell like cereal whenever the scent of toasted oats drifts through the city.

As lemon trees blossom along Italy's Amalfi coast and on the island of Sicily, a citrusy scent fills the air.

Sound . . .

Many animals communicate by sound waves that travel through the air.

When elephants make noise, they send out a sound that nearby elephants can hear with their ears. They also send sound waves that move through the ground. Elephants farther away can sense the waves with their feet.

Remember that volcanic eruption in the Sunda Strait? Krakatoa produced explosions so powerful they could be heard 3,000 miles or 4,800 kilometers away, making it the loudest sound in recorded history.

We need clean air for so many things, but air pollution is one of the biggest challenges we face.

All those particles that go up, up, up are changing the air around us, and that affects our planet and our health.
But we can all do something about it.

How can *you* help keep the air clean?

Ride your bike or walk whenever you can.

Use less energy—turn off lights and TVs when leaving rooms.

Plant trees that will help clean and cool the air.

Know your fire safety and help prevent wildfires.

Start an eco-friendly campaign in your neighborhood.

Ask your school or local organizations to invite scientists to talk about clean air.

Ask questions and learn about climate and weather.

Stay informed and share what you learn with others!

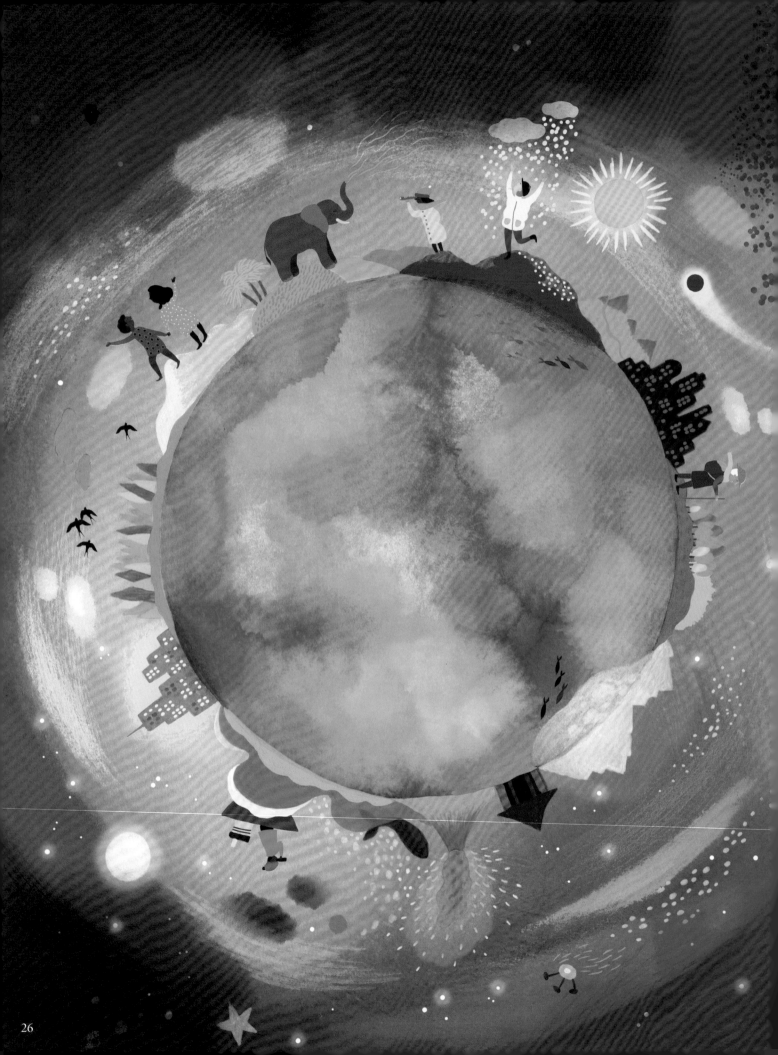

Everything is connected, and we all share the same air.

AUTHOR'S NOTE

The topic of AIR seems so simple but is really quite complex. In writing this book, I came across so many related topics and had to decide which ones fit this story best. Ultimately, I wanted to show the way air moves and how what we put into it in one place ends up in other places too.

While this book was being illustrated, dust from the Sahara desert moved through the air across the Atlantic Ocean to the Caribbean and parts of the United States, as it does nearly every year. Around the same time, smoke from wildfires in Canada traveled down the eastern part of the United States. It created dangerous conditions in the affected areas and New York City recorded the worst air quality in its history. Smoke particles formed a thick haze, and people witnessed unusually vivid orange and red daytime skies as the particles absorbed and bent light in different ways. Those smoke particles eventually traveled all the way across the Atlantic Ocean to Europe. We don't always see what goes into the air and where it travels, but these are visual reminders of how it happens.

There's a lot about air that I couldn't fit in this book, but the good news is that you can research and learn even more. You might want to read about the greenhouse effect, which has to do with warming the surface of the Earth. Or you might want to investigate the way bands of strong winds called jet streams blow across the globe from west to east, affecting the weather as well as air travel. And it's especially important to learn more about climate change and how these shifts in temperature and weather patterns over time affect our planet. While this book talks about the air *outside*, indoor air quality and energy-efficient buildings are also key pieces of the discussion.

You can also check out some really cool ways NASA is measuring factors that affect our air and our climate. An instrument named AIRS (Atmospheric Infrared Sounder) collects data from the Earth's atmosphere to improve weather forecasting, analyze air pollution, and support climate research. And EMIT (Earth Surface Mineral Dust Source Investigation) will help scientists understand how dust affects climate, and it can also monitor methane gases.

Understanding the science of air helps us make informed decisions in our daily lives. If you're curious about science, there are so many wonderful books, videos, websites, and experiments that will help you learn more and understand our world a little better.

We all share the same air, so let's work together to share the responsibility of keeping it clean!

GLOSSARY

GASES are substances that are not liquids or solids. They have no fixed shape but can expand to the shape of a container.

PARTICLES are extremely small pieces of matter.

Volcanic ASH is a mixture of rock fragments, minerals, and volcanic glass particles.

HAZE is the dust, smoke, and mist in the air that makes it hard to see.

We call the air surrounding the Earth the ATMOSPHERE.

COAL is a dark-colored rock that's found in the Earth and can be burned for fuel and electricity.

Things that pollute our air, water, or land are called POLLUTANTS.

Fog is like a cloud close to the ground. A DENSE FOG is hard to see through because it is so thick.

When something is burned, a fine black powder called SOOT is produced. It's what gives smoke its color.

SELECTED BIBLIOGRAPHY

"Atmosphere." *NASA Climate Kids*, NASA, https://climatekids.nasa.gov/menu/atmosphere/.

Blakemore, Erin. "The Great Smog of London Woke the World to the Dangers of Coal." *National Geographic*, National Geographic, 5 Dec. 2022, https://www.nationalgeographic.com/history/article/great-smog-of-london-1952-coal-air-pollution-environmental-disaster.

Burns, Ken, director. *The Dust Bowl - A Film By Ken Burns*. PBS, 2012.

Cooper, Michael L. *Dust to Eat: Drought and Depression in the 1930s.* Clarion Books, 2004.

"Domestic Fires Blamed For Half Britain's Smoke." *The Times*, 3 Dec. 1953, p. 5.

"The Great Java Eruption of 1883." *Birmingham Daily Post,* 30 Oct. 1888, p. 3.

"The Great Smog of 1952." *Met Office*, https://www.metoffice.gov.uk/weather/learn-about/weather/case-studies/great-smog.

Harpp, Karen. "How Do Volcanoes Affect World Climate?" *Scientific American*, Scientific American, 4 Oct. 2005, https://www.scientificamerican.com/article/how-do-volcanoes-affect-w/.

Lorentz, Pare, et al. *The Plow That Broke the Plains.* Resettlement Administration, 1960.

National Geographic Science of Everything: How Things Work In Our World. National Geographic, 2013.

Thomas, Isabel. *This Book Will (Help) Cool the Climate: 50 Ways to Cut Pollution and Protect Our Planet!* Random House, 2018.

Voiland, Adam. "A Clearer View of Hazy Skies." *NASA,* NASA, 24 June 2014, https://earthobservatory.nasa.gov/features/AirQuality.

Zimmerman, Dwight Jon, and Simon Winchester. *The Day the World Exploded: The Earthshaking Catastrophe at Krakatoa.* Collins, 2008.

INDEX

air, 2, 3, 5, 7, 11, 13, 14, 15, 16, 18, 19, 20, 22, 23, 24, 25, 27, 28, 29
Amalfi coast, 19
animals, 20
ash, 5, 6, 29
Atlantic Ocean, 10, 28
atmosphere, 6, 28, 29
bike, 25
birds, 17
Black Sunday, 10
British Parliament, 15
Buffalo, New York, 18
Canada, 28
Caribbean, 28
cereal, 18
Clean Air Act of 1956, 15
climate, 25, 28
coal, 13, 15, 29
dust, 8, 10, 11, 28, 29
Earth, 5, 6, 7, 28, 29
elephants, 20
energy, 15, 25, 28
environment, 16
Europe, 28
factories, 13, 15
fog, 14, 29
gases, 3, 6, 28, 29
greenhouse effect, 28
guano, 17
haze, 5, 28, 29
health, 23
homes, 13, 15
humans, 11
Italy, 19
Krakatoa, 4, 21
London, 13
New York City, 28
oats, 18
particles, 3, 9, 23, 28, 29
planet, 23, 28
plants, 9
pollutants, 13, 29
pollution, 15, 22, 28
rain, 9
Sahara, 28
science, 28
scientists, 7, 17, 25, 28
Sicily, 19
smog, 13, 14
smoke, 15, 28, 29
soil, 9, 17
soot, 14, 29
sound, 20, 21
storms, 8, 10, 11
Sunda Strait 4, 21
trees, 11, 19, 25
United States, 8, 11, 28
volcano, 4, 5
water, 11, 29
waves, 20
winds, 9, 17, 28

A *Books for a Better Earth*™ Title
The Books for a Better Earth™ collection is designed to inspire
young people to become active, knowledgeable participants in
caring for the planet they live on.
Focusing on solutions to climate change challenges and human
environmental impacts, the collection looks at how
scientists, activists, and young leaders are working
to safeguard Earth's future.

For Janet, who is braver than brave —D. R.

For my son, Martin —M. C.

ACKNOWLEDGMENTS

Thank you to Kellie DuBay Gillis, Melody Von Smith, and Kerri Zawadzki—experts in their fields
who have all been so generous with their time. And as always, the librarians at the Library of Congress
came through with exactly what I needed to find. Thank you! —D. R.

The publisher thanks Róisín Commane, PhD,
Department of Earth and Environmental Sciences,
Lamont-Doherty Earth Observatory, Columbia University;
and Matthew Blodgett, PhD.

Text copyright © 2025 by Deanna Romito
Illustrations copyright © 2025 by Mariona Cabassa
All Rights Reserved
HOLIDAY HOUSE is registered in the U.S. Patent and Trademark Office
Printed and bound in December 2024 at C&C Offset, Shenzhen, China.
This book was printed on FSC®-certified text paper.
The artwork was created with watercolors, gouache, and crayons.
www.holidayhouse.com
First Edition
1 3 5 7 9 10 8 6 4 2

Library of Congress Catalog-in-Publication Data

Names: Romito, Dee, author. | Cabassa, Mariona, illustrator.
Title: The air we share : a pollution problem and finding ways to
fix it / by Dee Romito ; illustrated by Mariona Cabassa.
Description: First edition. | New York : Holiday House, [2025] | Includes bibliographical references and index.
Audience: Ages 6–9 | Audience: Grades 2–3 | Summary: "An introduction to the atmosphere,
how air pollution affects the environment, and what we can do to keep the air clean"–Provided by publisher.
Identifiers: LCCN 2023050669 | ISBN 9780823455003 (hardcover)
Subjects: LCSH: Air–Pollution–Juvenile literature. | Air quality–Juvenile literature. | Air–Juvenile literature.
Classification: LCC TD883.13 .R66 2025 | DDC 363.739/2–dc23/eng/20240322
LC record available at https://lccn.loc.gov/2023050669

ISBN: 978-0-8234-5500-3 (hardcover)

EU Authorized Representative: HackettFlynn Ltd., 36 Cloch Choirneal,
Balrothery, Co. Dublin, K32 C942, Ireland. EU@walkerpublishinggroup.com